INSIDE SCIENCE

MOON
AND
TIDES

Rob Lang

CONTENTS

THE MOON

On July 21, 1969, Neil Armstrong became the first person to walk on the moon. The surface was like powdered charcoal. His first step left a crater about 12 inches (30.48 centimeters) deep. What is a moon? Does every planet have one? How does a planet get one? Does having a moon near your planet have an effect on you?

The moon is the earth's only natural **satellite**. When you see it at night, it is the brightest object in the sky. But did you know that we can only see the moon because it reflects sunlight?

Moon

Light from the sun

Reflected sunlight from the moon

Earth

Sun

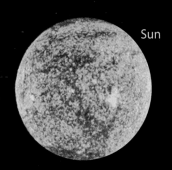

Earth's moon doesn't just sit in one place in space. It orbits the earth. Its orbit follows an **elliptical** path around our planet.

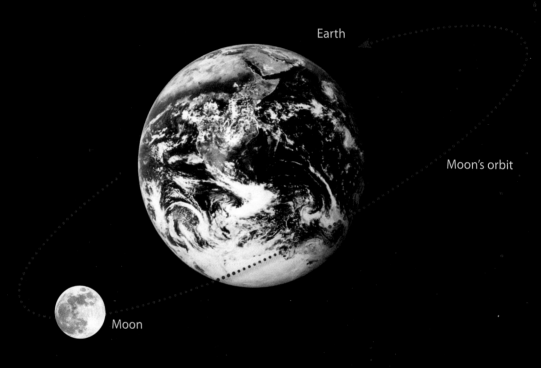

Earth

Moon's orbit

Moon

Many other planets have moons orbiting them, too. Some planets have more than one. There are two moons orbiting Mars, for example. They are shaped like potatoes!

Phobos

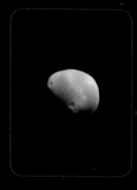

Deimos

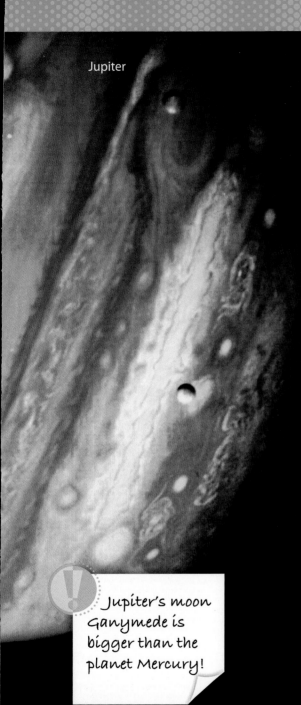

Jupiter

The planet in our **solar** system with the most moons is Jupiter. It has 63 moons. Saturn is the next, with 60 moons. Imagine what it would look like at night if the earth had 60 moons—or even two!

Jupiter also has the largest moon in our solar system—Ganymede.

Jupiter's moon Ganymede is bigger than the planet Mercury!

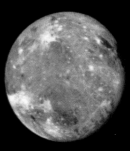

Ganymede

The Big Whack Theory

Many scientists believe that our moon may have formed when a planet-sized object crashed into the earth about 4.5 billion years ago. Here's the **theory**.

1

A planet-sized object crashes into the earth, throwing up a huge amount of dust and rock.

2

This material is captured by the earth's gravity and begins orbiting the earth.

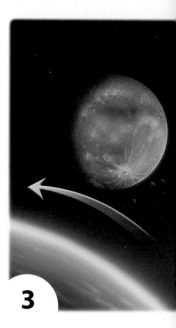

3

As it orbits, the moon forms.

If this is true, it helps to explain why the moon's surface is covered with **craters**. **Meteoroids** have been smashing into the moon's surface for millions and millions of years, creating craters—and there is no weather on the moon to blow or wash them away.

There is also a theory that the earth once had three moons.

The far side of the moon, as seen from the *Galileo* spacecraft

The next time you're outside on a clear night, look at the light and dark patches on the moon. What do you think they are?

The dark parts are smooth areas where the magma—molten rock—flowed out while the moon was "young."

The light parts are mountains that weren't covered by magma.

The moon is hit by more meteoroids than the earth. This is because the moon doesn't have a thick **atmosphere** to vaporize—burn up— meteoroids before they reach the surface.

Scientists used to think that the moon didn't have an atmosphere. Now they know that it has a very thin one, called an exosphere. Most of the moon's exosphere arrives by way of the solar wind, a stream of gas particles coming from the sun. Part of the moon's exosphere may also be gas leaking out of the moon's rocks. The rest may be **molecules** of vaporized rock from meteoroid strikes.

The rocket exhaust from each *Apollo* landing temporarily doubled the moon's exosphere.

The lunar exosphere

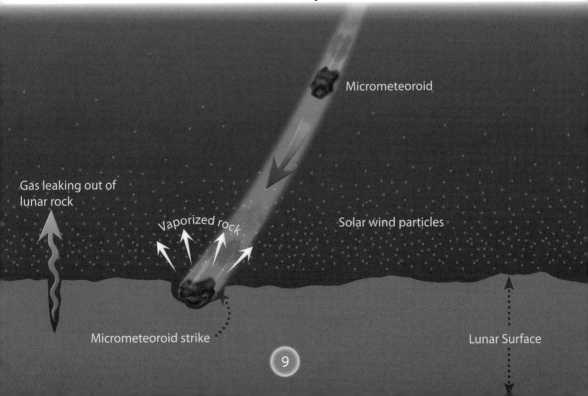

Micrometeoroid

Gas leaking out of lunar rock

Vaporized rock

Solar wind particles

Micrometeoroid strike

Lunar Surface

How Far Away Is the Moon?

If you wanted to visit the moon to take a closer look at its craters, how far would you need to travel? Scientists have measured the distance between the earth and the moon using a laser beam. The beam bounces off reflectors on the moon and returns to the earth. How did reflectors get on the moon? Astronauts put them there!

A Laser Ranging Retro Reflector deployed by the *Apollo 14* astronauts

Moon

A laser beam sent from earth bounces off a reflector on the moon.

Earth

Using the reflectors on the moon, scientists have discovered that the moon is slowly leaving the earth's orbit. Imagine a night sky without the moon! Don't worry, though—it won't happen in your lifetime. The moon is only moving about 1.5 inches (3.81 centimeters) away from the earth each year. You could be growing faster than that!

To get to the moon, you'd need to travel about 238,900 miles (384,472 kilometers). That's the same as making 80 trips across the United States!

10

Temperatures on the Moon

The lunar sky

Imagine that you did travel to the moon. What would you see when you got there? Apart from the sun, you'd see a dark sky—and it would be either very hot or very cold, depending on whether you were standing in a shadow or not.

You'd need to wear a temperature-controlled spacesuit on the moon. Out of the sunlight, it gets as cold as -400°F (-240°C). In direct sunlight, the temperature gets as hot as 260°F (126.7°C).

How cold and hot is this? At sea level on the earth, freshwater freezes at 32°F (0°C) and boils at 212°F (100°C), so don't take your spacesuit off if you visit the moon!

Did you know that the sky isn't blue when looked at from the moon? You need an atmosphere like the earth's in order to see a blue sky.

Does the Moon Revolve on Its Axis?

The moon orbits around the earth, but does it revolve on its axis while it is doing so, the way the earth revolves on its axis? After all, we always see the same side of the moon, so it doesn't *look* as if it's revolving.

To find out, you will need:

- a big ball and a little ball

- a marker

In the time that it takes for the moon to revolve on its axis once, the earth revolves on its axis over 27 times. A day on the moon, measured in earth time, is 27 days, 7 hours, and 43 minutes. (Of course, measured in moon time, it's just one lunar day!)

What to do:

1. Imagine that the bigger ball is the earth. Put it on the floor.

2. Imagine that the smaller ball is the moon. Put an X on one side of it.

3. Now orbit the moon around the earth, *always* keeping the side with the X on it facing the earth.

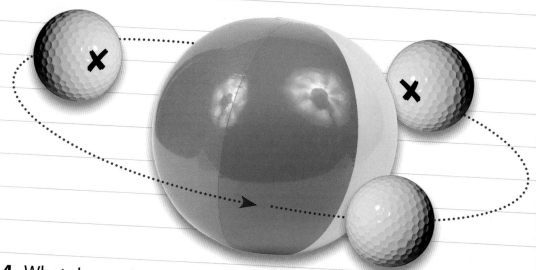

4. What do you have to do to the moon to make sure that the same side—the side with the X on it—always faces the earth?

Show a friend:

Use your model to prove to a friend that the moon does in fact revolve on its axis.

Phases of the Moon

As the moon orbits the earth, the moon's shape seems to change. This is because the side of the moon that faces the earth is exposed to different amounts of sunlight. This change follows a pattern called "the phases of the moon."

Just like the sun, the moon rises in the east and sets in the west. Each night, the moon rises about 50 minutes later than it did the night before.

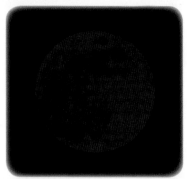

New moon

First quarter

Full moon

Last quarter

ECLIPSES

Sometimes, the earth gets between the sun and the moon and casts a shadow over the moon. When the earth stops sunlight from reaching the moon, we call this a **lunar** eclipse.

Moon   Earth   Sun

At other times, the moon passes between the earth and the sun. The moon isn't big enough to block out sunlight all over the earth. It just casts a dark band of shadow across part of the earth. This is called a solar eclipse.

Earth   Moon   Sun

A solar eclipse
seen from space

Warning! A solar eclipse is a strange thing to experience, but you must never look right at one. You could damage your eyes.

GRAVITY

What makes the earth orbit the sun? What makes the moon orbit the earth? Why don't they both just drift off in different directions? It's all because of gravity—a force that pulls objects together.

Weight is a way to describe the force of gravity pulling objects down. This is how heavy they feel. On the moon, the force of gravity is only one-sixth as strong as on earth (because the moon is smaller and lighter). This is why an astronaut weighs six times less on the moon than on earth. There's still just as much astronaut but, because there is less gravity, his or her body weighs less on the moon!

TIDES

The water level of every sea and lake on the earth slowly rises and falls, day after day. We call this change "the tide." Right now, as you're reading this page, it's high tide somewhere on the earth. Somewhere else, it's low tide.

Low tide—
the water has gone out.

What makes water levels change like this? The answer is gravity.

High tide—
the water has come in.

Even though the moon's gravity is weaker than the earth's, the moon is still close enough to the earth to affect large bodies of water on it by pulling on them, creating tides.

Did you know that the moon pulls on earth's rocks, too? In fact, there are tides in the solid parts of our planet. You can't see them, because the effect is so small, but scientists can measure this pull with very **sensitive** instruments.

High and low tides

Low tide

Water is pulled by the moon's gravity.

Water is left behind as a bulge.

High tide

Moon

Earth

High tide

The earth is pulled a little bit toward the moon.

Low tide

As the earth rotates, different parts of the earth are closer to the moon at different times. Where water is pulled toward the moon, it is high tide. Where it isn't, it is low tide. It is high tide at the same time on the other side of the earth because the water there is left behind as the earth is pulled away by the moon's gravity.

There are usually two high tides and two low tides every day, but they do not occur exactly 12 hours apart in a period of 24 hours, as you might expect. If one high tide is at 6 a.m., for example, the next high tide will be 12 hours and 25 minutes later, at 6:25.

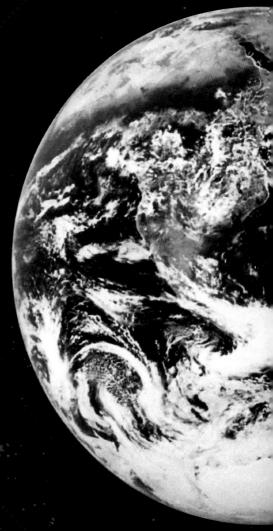

One full rotation
of the earth

The moon has moved on!

12°

The earth rotates on its axis once every 24 hours. In the same time, the moon moves through its orbit 12°. Catching up to the moon accounts for each tide coming 25 minutes later than you'd expect!

High and low tides are different in different places on the earth. This is because the moon has a tilted, elliptical orbit.

In parts of Hawaii, the tidal range is only 2 feet (0.6 metres).

High tide in Hawaii

Low tide in Hawaii

The shape of the sea floor along a coastline has a big effect on the height of tides, too.

When the sun and the moon are both pulling together, tides are extreme.

In the Bay of Fundy, in Canada, the difference between high tide and low tide can be as large as 50 feet (15.24 meters)!

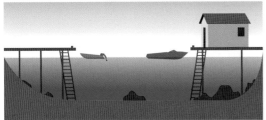

High tide in the Bay of Fundy

Low tide in the Bay of Fundy

Have you ever made a sandcastle at the beach? Did the high tide wash it away? Isn't it strange to think that it was the pull of the moon's gravity that was responsible for washing your sandcastle away? If you think that's amazing, just think how different things would be if you lived on a planet that didn't have a moon—or if your planet had 60 moons!

GLOSSARY

atmosphere—the gases surrounding a planet

craters—holes in the ground made by meteoroids or volcanoes

elliptical—oval-shaped

ion—an atom with an electric charge

lunar—to do with the moon

meteoroids—a piece of space material entering a planet's atmosphere or moon's exosphere (the pieces that survive impact are called meteorites— tiny meteoroids are called micrometeoroids)

molecules—atoms bonded together

satellite—an object orbiting another

sensitive—able to detect a very small change

shielded—protected

solar—to do with the sun

theory—an idea that can be scientifically tested